MATRIX INVERSIONS VIA JIBUNOH'S DETERMINANTS AND EXACT SOLUTIONS OF K X K SYSTEMS OF LINEAR EQUATIONS

A monograph on research discovery

C. C. JIBUNOH
B.Sc, M.Ed, M.Sc, D.Sc, MNMS, MNSA

All Rights Reserved
No part of this publication may be reproduced, by any means,
Be it electronic, mechanical or otherwise, without the
Prior permission in writing from the author

ISBN: 978-1494291662

First Published in 2010

© C. C. JIBUNOH
Email: chafachid@gmail.com GSM: +2347063860446, +2348053336496
Department of Mathematics and Statistics
Delta State Polytechnic Ogwashi-Uku
Nigeria

DEDICATION

TO MY SONS CHIKWADOM AND CHICHEBEM WHO ARE YOUNGER AND ZEALOUS MATHEMATICIANS

PREFACE

My previous work on Determinants, namely 'Jibunoh's method for evaluating the determinant of an n x n matrix; a monograph on research discovery' focused on a new and simple approach for evaluating the determinant of any n x n matrix, by reduction to echelon form. This approach demonstrated superiority (in terms of more simplicity) over the traditional method of cofactors which is tedious to apply when n is large. The contents of 'Jibunoh's determinants' gained popularity after the initial presentation to the Nigerian Mathematical Society (NMS) in the conference of June 2009, at Ilorin, Nigeria.

In 2010, it became necessary to extend the concept of Jibunoh's determinants to the quick and exact evaluation of inverses of k x k matrices, no matter how large the value of k. This naturally resulted in obtaining the exact numerical solutions of any corresponding k x k systems of linear equations. The work was also a product of research which was first presented in a workshop at the National Mathematical Centre, Abuja, in June 2010, where it was received with great enthusiasm. The Abuja presentation formed the contents of the current monograph which now bears the title *'Matrix inversions via Jibunoh's determinants and exact solutions of k x k systems of linear equations'*

I expect that the monograph should be a companion to my first monograph of 2009 which has circulated widely and which should be studied in order to grasp the simple but technical aspect of matrix inversions via Jibunoh's determinants. Nevertheless, the current monograph has been made to be relatively simple and self contained.

As usual, I recommend the monograph not only to mathematicians or statisticians per se, but to engineers, technologists, social scientists, business administrators, accountants, computer scientists etc, including students in these fields who are in the secondary and tertiary institutions. They will appreciate the simplicity of the new method of matrix inversions and the beauty of obtaining the exact solutions of k x k systems of linear systems.

I must thank the National Mathematical Centre, Abuja, for creating a workshop in 2010, during which I gave my maiden presentation. I should not fail to thank my colleagues in the Nigerian Mathematical Society and my children, friends and relatives for their moral support and encouragement.

Dr C.C. Jibunoh
Ogwashi-Uku, Nigeria
25th November, 2010

ARRANGEMENT OF SECTIONS

	Page
Abstract	1
1. Introduction	1
2. Solution of the linear system by Jibunoh's Method and backward vector substitutions(b.v.s)	2
3. Converting the fractional or decimal entries of the equation matrix to integral entries	4
4. Simplifications in the case when the entries of the system are complex numbers	5
5. Solutions of complex systems	7
6. Numerical Applications (Examples)	8
7. Conclusion	28
8. Exercises	29
References	37
Index	38

MATRIX INVERSIONS VIA JIBUNOH'S DETERMINANTS AND EXACT SOLUTIONS OF K × K SYSTEMS OF LINEAR EQUATIONS

BY

CHAFA C. JIBUNOH
(chafa_ chid @ yahoo.com, +2348053336496)
Department of Mathematics and Statistics
Delta State Polytechnic, Ogwashi-Uku, Nigeria

As presented in a workshop at the National Mathematical Centre, Abuja
June, 2010

Abstract: *A simple and systematic procedure for solving any k × k system of linear equations is developed in this paper. The determinant of the equation matrix is first found using Jibunoh's method. Then the matrix is inverted by applying the defined backward vector substitutions (bvs). The reciprocal of the positive value of the determinant, if the matrix is real, is taken as a factor of the inverse matrix. The complex matrix is similarly inverted to obtain what is defined as either the Analytical or Empirical inverse. The entries of any inverse matrix (real or complex) are mainly integers, without the scalar-factor multiplying the matrix. This makes the inverse matrix exact and more accurate than decimal representations obtained by computer evaluations. For any system of equations, therefore, three quantities are obtained simultaneously, namely, the determinant of the equation matrix, the inverse of the matrix and the solution of the system. The production of these quantities simultaneously, is new in the literature. By these procedures, any linear systems of equations of dimensions k can be solved easily and accurately, as $k \to \infty$.*

1. Introduction

The traditional approach to solving a k × k system of linear equations could be by Gaussian elimination or by direct inversion of the equation matrix. Gaussian elimination involves backward scalar substitutions which in some cases are prone to error if we fail to apply pivoting strategies. For example in Burden and Faires [1] p.338, the system

$$0.003x_1 + 59.14x_2 = 59.17 \qquad (1.1)$$
$$5.291x_1 - 6.13x_2 = 46.78$$

could not be solved accurately by Gaussian elimination without the so-called maximal column pivoting. Iterative methods in the literature e.g. Gauss-Seidel and Jacobi, also have problems of accuracy.

Matrix inversion by use of classical adjoint and cofactors involve the evaluation of the determinant of the equation matrix such that if the equation is

A x = b \qquad (1.2)

where **A** is the k × k matrix and **b** is a k- component column vector, then

$$x = \frac{1}{\det A}(\mathrm{adj}\, A)b, \qquad (1.3)$$

provided det A $\neq$ 0, Again by use of co-factors, det **A** is tedious to find if k is large. Hence we often resort to computer evaluation which in most cases gives the entries of inverse matrix to some decimal places approximately, and not exact valves, and does not produce det **A**, **A**$^{-1}$ and solution of the system simultaneously.

In this paper, we shall adopt Jibunoh's method [2] for evaluating the determinant of any n×n matrix and apply it easily to obtaining the inverse of the equation matrix which ultimately leads to the solution of the system. Three quantities are thus obtained simultaneously, namely, the determinant of the equation matrix, the inverse of the matrix and the solution of the system. The production of these quantities simultaneously, is new in the literature. Because det **A** is always found easily by Jibunoh's method, and provided it is non zero, its positive reciprocal, in the case of a real matrix, is made to be a multiplicative factor of the entries of the inverse matrix in the form;

$$\mathbf{A}^{-1} = \frac{1}{+\det A}\begin{pmatrix} y_{11} & y_{12} & \cdots & y_{1k} \\ y_{21} & y_{22} & \cdots & y_{2k} \\ - & - & - & - \\ - & - & - & - \\ y_{k1} & y_{k2} & \cdots & y_{kk} \end{pmatrix} \qquad (1.4)$$

where y_{ij} are mainly integers if the original entries of A are integers. The complex matrix is inverted similarly with a different form of the multiplicative factor to give what is defined as either the Analytical or Empirical inverse. The form (1.4) gives the exact inverse of **A**, without approximations. The process leading to (1.4) is carried out by what we define as backward vector substitutions (bvs).

2. Solution of the linear system by Jibunoh's method and backward vector substitution (b.v.s)

Let the matrix equation be given by

Ax = b

as in (1.2).
Then,

x = A^{-1}b $\qquad (2.1)$

Now,

AA^{-1} = I $\qquad (2.2)$

Let the entries of **A⁻¹** be given by k row vectors denoted jointly as **y**, then
Ay = I (2.3)
where

$$\mathbf{y} = \begin{pmatrix} y_1 \\ y_2 \\ . \\ . \\ y_k \end{pmatrix}$$

and y_j are row vectors, each of k components.
By reducing the matrixes A and I on both sides of (2.3) to echelon form, using Jibunoh's method, we obtain an alternative form of (2.3)

i.e. **Cy = B** (2.4)

where **C** is upper triangular matrix, reduced from A and **B** is the corresponding lower triangular matrix reduced from I. Now writing (2.4) in the form;

$$C \begin{pmatrix} y_1 \\ y_2 \\ . \\ . \\ y_k \end{pmatrix} = B,$$ (2.5)

the row vectors $y_1, y_2, \ldots y_k$, are found by backward substitutions. Hence these substitutions are defined as backward vector substitutions (b.v.s). From Jibunoh's method [2], the reduction of A to the echelon matrix **C**, leads to obtaining det **A**, the determinant of **A**.

By b.v.s, from (2.5), we must first obtain the row vector y_k.
Suppose

$$m y_k = (\alpha_{k1}, \alpha_{k2}, \ldots \alpha_{kk})$$ (2.6)

where m is a constant number.
Then to express (2.6) in terms of the positive value of det **A**, we write

$$\det A\, m\, y_k = \det A\, (\alpha_{k1}, \alpha_{k2} \ldots \alpha_{kk}), \quad \text{if } \det A > 0$$ (2.7)

or

$$-\det A\, m\, y_k = -\det A\, (\alpha_{k1}, \alpha_{k2} \ldots \alpha_{kk}), \quad \text{if } \det A < 0$$ (2.8)

In either case of (2.7) or (2.8)

$$\det A\, y_k = \frac{\det A}{m} (\alpha_{k1}, \alpha_{k2}, \ldots \alpha_{kk}) = (y_{k1}, y_{k2}, \ldots y_{kk}), \text{ after multiplication.}$$ (2.9)

It often turns out that (y_{k1}, y_{k2}, ... y_{kk}) is a row vector with mainly integral components if the matrix A has integral entries.

Hence $y_k = \dfrac{1}{+\det A}(y_{k1}, y_{k2}, \ldots y_{kk})$ \hfill (2.10)

Using y_k, then further backward substitutions are made and manipulated to obtain all the row vectors, with the positive reciprocal of the determinant as a factor, as follows;

$$y_k = \dfrac{1}{+\det A}(y_{k1} \quad y_{k2}, \ldots \ldots \quad y_{kk})$$

$$y_{k-1} = \dfrac{1}{+\det A}(y_{k-1,1}, \quad y_{k-1,2} \ldots y_{k-1,k}) \qquad (2.11)$$

....

....

$$y_1 = \dfrac{1}{+\det A}(y_{11} \quad y_{12} \quad \ldots \quad y_{1k})$$

Hence

$$A^{-1} = \dfrac{1}{+\det A}\begin{pmatrix} y_{11} & y_{12} & \ldots y_{1k} \\ y_{21} & y_{22} & \ldots y_{2k} \\ -- & - & -- \\ -- & - & -- \\ y_{k1} & y_{k2} & \ldots y_{kk} \end{pmatrix}$$

as given in (1.4) where y_{ij} are mainly integers if the original matrix A has integral entries. Having obtained the inverse of A, then the system (1.2) is solved using (2.1). The simple procedures of obtaining the row vectors (2.11) are shown in the examples, in section 6.

The work in Jibunoh [2], should be consulted for more details of Jibunoh's method for evaluating the determinant of a k × k matrix.

3. Converting the fractional or decimal entries of the equation matrix to integral entries.

Suppose the equation matrix A has fractional or decimal entries which are real or complex. To simplify the process of finding the determinant or the inverse of A, we may convert the entries to integers.

Let $β_1$ $β_2$ ... $β_k$ be the scalars required to convert rows 1 to k respectively of A, to integral entries. Then writing,

$$B = \begin{pmatrix} \beta_1 & 0 & \text{---------} & 0 \\ 0 & \beta_2 & \text{---------} & 0 \\ -- & --- & \text{---------} & -- \\ 0 & 0 & \text{---------} & \beta_k \end{pmatrix}, \quad (3.1)$$

as a diagonal matrix, we have;

$$BA = A^* \quad (3.2)$$

where A* is the new matrix with integral entries. Denoting detA as $|A|$ we deduce

$$|B||A| = |A^*| \quad (3.3)$$

or

$$\beta_1 \beta_2 \ldots \beta_k |A| = |A^*|$$

Therefore

$$|A| = \frac{|A^*|}{\beta_1 \beta_2 \ldots \beta_k} \quad (3.4)$$

Now from (3.2), we have;

$$A^{-1} = A^{*-1} B \quad (3.5)$$

If A is complex then for ease of computation A^{*-1} is put in the form of (4.9), below.

4. Simplifications in the case when the entries of the system are complex numbers

Let $C_1 + i C_2$ and $\alpha_1 + i\alpha_2$ be two complex numbers: Traditionally,

$$(C_1 + i C_2)(\alpha_1 + i\alpha_2) = (C_1\alpha_1 - C_2\alpha_2) + i(C_1\alpha_2 + C_2\alpha_1) \quad (4.1)$$

By a simplification process, these products and sums can be obtained easily by writing the LHS of (4.1) as

$$(C_1, C_2) \begin{bmatrix} \alpha_1 & \alpha_2 \\ -\alpha_2 & \alpha_1 \end{bmatrix} \quad (4.2)$$

where the first complex number is turned to a (1x2) row vector and the second complex number is turned to the given 2x 2 matrix.

The product of (4.2) leads to the row vector;

$$(C_1\alpha_1 - C_2\alpha_2, \quad C_1\alpha_2 + C_2\alpha_1)$$

which is written down immediately as

$$(C_1\alpha_1 - C_2\alpha_2) + (C_1\alpha_2 + C_2\alpha_1)i$$

to give the RHS of (4.1)

If a scalar complex number is to multiply a complex row vector as, for example, the case

$$(\alpha_1 + i\alpha_2)(C_{11} + iC_{21},\ C_{12} + iC_{22},\ \ldots,\ C_{1k} + iC_{2k}) \quad (4.3)$$

it is preferable to transpose the row vector and multiply as in (4.2), i.e

$$\begin{bmatrix} C_{11} & C_{21} \\ C_{12} & C_{22} \\ ----- \\ ------ \\ C_{1k} & C_{2k} \end{bmatrix} \begin{bmatrix} \alpha_1 & \alpha_2 \\ -\alpha_2 & \alpha_1 \end{bmatrix} \quad (4.4)$$

which at once gives the complex vector

$$\begin{pmatrix} (C_{11}\alpha_1 - C_{21}\alpha_2) & + (C_{11}\alpha_2 + C_{21}\alpha_1)i \\ (C_{12}\alpha_1 - C_{22}\alpha_2) & + (C_{12}\alpha_2 + C_{22}\alpha_1)i \\ ------------------- \\ ------------------- \\ (C_{1k}\alpha_1 - C_{2k}\alpha_2) & + (C_{1k}\alpha_2 + C_{2k}\alpha_1)i \end{pmatrix}^T \quad (4.5)$$

In the case where the product of several complex numbers are required e.g.

$$(C_1 + iC_2)(a_1 + ia_2)(b_1 + ib_2)(d_1 + id_2) \quad (4.6)$$

we fix one of the numbers as a row vector and convert the other numbers to 2 x 2 matrices Hence (4.6) can be obtained as;

$$(C_1, C_2)\begin{bmatrix} a_1 & a_2 \\ -a_2 & a_1 \end{bmatrix}\begin{bmatrix} b_1 & b_2 \\ -b_2 & b_1 \end{bmatrix}\begin{bmatrix} d_1 & d_2 \\ -d_2 & d_1 \end{bmatrix} \quad (4.7)$$

The product of (4.7) gives a row vector which shall be written down at once in the complex format. In section 2, the positive reciprocal of det A is usually made to be a factor of the inverse matrix when A has real entries. When A has complex entries, it is simpler to use the reciprocal of the product of the diagonal elements of the echelon matrix of A, as a factor.

Where any k x k complex matrix A, is such that k is too large or the entries of A are mainly decimal numbers as distinct from integers, manual evaluation of A^{-1} may be tedious. Therefore after obtaining the determinant of A by Jibunoh's method, an alternative approach to finding A^{-1} is as follows:

Suppose; $A = A_1 + iA_2$ \quad (4.8)

where A_1 and A_2 are the real and imaginary parts of A.

Let

$A^{-1} = V_1 + i V_2$ (4.9)

where V_1 and V_2 are k × k matrices which are to be found. To find V_1 and V_2, we note that
$AA^{-1} = I$

Hence
$A_1V_1 - A_2V_2 = I$
$A_2V_1 + A_1V_2 = 0$ (4.10)

where I and O are k × k matrices respectively
Then

$$\begin{pmatrix} V_1 \\ --- \\ V_2 \end{pmatrix} = \begin{pmatrix} A_1 & \vdots & -A_2 \\ -- & -\vdots & -- \\ A_2 & \vdots & A_1 \end{pmatrix}^{-1} \begin{pmatrix} I \\ --- \\ O \end{pmatrix}$$ (4.11)

From (4.11), V_1 and V_2 are then found by automatic computation using, for example, the Microsoft Excel Package. We shall define the form of the inverse given by (4.9) as the *Empirical inverse* of A, while the form which is not split into real and imaginary matrices shall be called *Analytical inverse* of A. The Analytical can be changed to the Empirical inverse by a simple rationalization.

5. Solutions of Complex Systems

Suppose a complex linear system is represented by (1.2) i.e.
Ax = b
where A is, in general, a complex matrix and b is a complex column vector.
Then by (2.1),
x = A⁻¹b
we let **A⁻¹** be the Empirical Inverse of A which is given by (4.9)
i.e.
A⁻¹ = V₁ + iV₂
and
b = $b_1 + ib_2$ (5.1)

where b_1 and b_2 are the real and imaginary parts of the column vector **b**.
It follows that
x = A⁻¹b = $(V_1 + iV_2)(b_1 + ib_2)$ (5.2)
Hence
x = $(V_1b_1 - V_2b_2) + (V_1b_2 + V_2b_1)i$, (5.3)

which is transposed to a column vector.
Thus the solution of the system is easily obtained by using the **A⁻¹**(Empirical).
Observe from (5.3), that;

Re $\mathbf{x} = V_1 b_1 - V_2 b_2$

and (5.4)

Im $\mathbf{x} = V_1 b_2 + V_2 b_1$

6. NUMERICAL APPLICATIONS (EXAMPLES)

Example I

Using Jibunoh's method find the determinant and inverse of the matrix

$$A = \begin{pmatrix} 2 & 3 \\ 5 & 6 \end{pmatrix} \tag{6.1}$$

This is a 2×2 matrix which can be handled easily by the traditional method of cofactors. But we need to see the workings of Jibunoh's method which extends easily to large scale matrices. We shall find both the determinant and inverse of the matrix, simultaneously.

By (2.3) we write

$$\begin{pmatrix} 2 & 3 \\ 5 & 6 \end{pmatrix} \begin{pmatrix} y_1 \\ y_2 \end{pmatrix} = \begin{pmatrix} 1 & 0 \\ 0 & 1 \end{pmatrix}, \text{ where } y_1 \text{ and } y_2 \text{ are row vectors.} \tag{6.2}$$

To reduce the equation matrix A, on the LHS of (6.2), to echelon form we temporarily drop the vector $\begin{pmatrix} y_1 \\ y_2 \end{pmatrix}$ and apply the format of reduction to echelon form given in [2], as follows;

2	3	1	0	5
5	6	0	1	(2)

(2)

2	3	1	0	
0	-3	-5	2	

(3)

As defined in [2], the number 2 in parentheses, under 5, is a lower multiplier. Now when the matrix A is reduced to echelon form as given above by;

$$\mathbf{C} = \begin{pmatrix} 2 & 3 \\ 0 & -3 \end{pmatrix}, \text{ then as proved in [2],}$$

$$\det \mathbf{A} = \frac{\text{product of diagonal elements of } \mathbf{C}}{\text{product of lower multipliers from each stage of the reduction to } \mathbf{C}} \tag{6.4}$$

Hence
$$\det A = \frac{2 \times -3}{2} = -3 \tag{6.5}$$

Re-introducing the vector $\begin{pmatrix} y_1 \\ y_2 \end{pmatrix}$ alongside the echelon matrix in (6.3) and equating to the matrix on the RHS, we have;

$$\begin{pmatrix} 2 & 3 \\ 0 & -3 \end{pmatrix} \begin{pmatrix} y_1 \\ y_2 \end{pmatrix} = \begin{pmatrix} 1 & 0 \\ -5 & 2 \end{pmatrix} \tag{6.6}$$

By backward vector substitutions (b.v.s), we have
$$-3 y_2 = (-5, \ 2) \tag{6.7}$$

Using (2.8), we multiply both sides of (6.7) by the determinant of **A**, which is − 3, to obtain

$$-3(-3)y_2 = -3(-5, \ 2)$$

i.e. $\ 3(3)y_2 = (15, \ -6)$

$3y_2 = (5, \ -2)$

$y_2 = \frac{1}{3}(5, \ -2)$

By bvs;

$2y_1 + 3y_2 = (1, \ 0) = \frac{1}{3}(3, \ 0)$, expressing the RHS in terms of the positive reciprocal of

det**A**

$2y_1 + \frac{1}{3}(15, \ -6) = \frac{1}{3}(3, \ 0)$

$2y_1 = \frac{1}{3}(-12, \ 6)$

$y_1 = \frac{1}{3}(-6, \ 3)$

Therefore

$$A^{-1} = \frac{1}{3} \begin{pmatrix} -6 & 3 \\ 5 & -2 \end{pmatrix}$$

det **A** = -3.

Example 2

Use Jibunoh's method to find the determinant and inverse of the matrix

$$A = \begin{pmatrix} 1 & 2 & 3 \\ 4 & 5 & 6 \\ 7 & 8 & 5 \end{pmatrix}$$

We proceed with the method of reduction to echelon form as explained in Example 1, but the reader should see the procedure in [2] where only the determinant of this matrix was found.

1	2	3	1	0	0	4	7
4	5	6	0	1	0	(1)	
7	8	5	0	0	1		(1)
1	2	3	1	0	0		
0	-3	-6	-4	1	0	2	
0	-6	-16	-7	0	1		(1)
1	2	3	1	0	0		
0	-3	-6	-4	1	0		
0	0	-4	1	-2	1		

$$\det \mathbf{A} = \frac{1 \times -3 \times -4}{1 \times 1 \times 1} = 12$$

Now,

$$\begin{pmatrix} 1 & 2 & 3 \\ 0 & -3 & -6 \\ 0 & 0 & -4 \end{pmatrix} \begin{pmatrix} y_1 \\ y_2 \\ y_3 \end{pmatrix} = \begin{pmatrix} 1 & 0 & 0 \\ -4 & 1 & 0 \\ 1 & -2 & 1 \end{pmatrix}$$

By bvs,

$-4y_3 = (1, -2, 1)$

$-4(12)y_3 = 12(1, -2, 1) = (12, -24, 12)$ since det A = 12

$12y_3 = (-3, 6, -3)$

$y_3 = \dfrac{1}{12}(-3, 6, -3)$

By further bvs;

$-3y_2 - 6y_3 = (-4, 1, 0) = \dfrac{1}{12}(-48, 12, 0)$

$-3y_2 + \dfrac{1}{12}(18, -36, 18) = \dfrac{1}{12}(-48, 12, 0)$, using -6 to multiply the known vector y_3

$y_2 = \dfrac{1}{12}(22, -16, 6)$

Also, by further bvs;

$y_1 + 2y_2 + 3y_3 = (1, 0, 0) = \dfrac{1}{12}(12, 0, 0)$

Hence $y_1 = \dfrac{1}{12}(-23, 14, -3)$

$$\mathbf{A^{-1}} = \dfrac{1}{12}\begin{pmatrix} -23 & 14 & -3 \\ 22 & -16 & 6 \\ -3 & 6 & -3 \end{pmatrix}$$

det **A** = 12

Example 3 (a)

Use Jibunoh's method to solve the system of equations

$$2x_1 \quad\quad\quad - x_3 = 4$$
$$3x_1 \quad\quad\quad + 2x_3 = 5$$
$$4x_1 - 3x_2 + 7x_3 = 6$$

Note: *In a system of linear equations we use $y_1, y_2, \ldots y_k$ to denote row vectors in backward substitutions when finding the inverse of the equation matrix and $x_1, x_2, \ldots x_k$ to denote the solutions of the system.*

We, therefore, write the given matrix equation as;

Ax = b

i.e.

$$\begin{pmatrix} 2 & 0 & -1 \\ 3 & 0 & 2 \\ 4 & -3 & 7 \end{pmatrix} \begin{pmatrix} x_1 \\ x_2 \\ x_3 \end{pmatrix} = \begin{pmatrix} 4 \\ 5 \\ 6 \end{pmatrix}$$

By the theory in [2] it is pointed out that since

$$\det \mathbf{A_2} = \begin{vmatrix} 2 & 0 \\ 3 & 0 \end{vmatrix} = 0,$$

we need to interchange two rows or columns so to make $\det \mathbf{A_2} \neq 0$. *In solving systems of equations or finding matrix inverses row interchanges only, are recommended, where necessary, due to the simplicity of the algebra.*

Hence interchanging the 2nd and 3rd rows we have the new matrix

in which $\det \mathbf{A_2} \neq 0$

We first obtain the determinant and inverse of this matrix before solving the system. Hence leaving out the RHS of the system for the time being we proceed with the following reductions, after also interchanging the 2nd and 3rd rows of the identity matrix, since we have interchanged the 2nd and 3rd rows of the matrix **A**.

2	0	-1	1	0	0	2	3
4	-3	7	0	0	1	(1)	
3	0	2	0	1	0		(2)
2	0	-1	1	0	0		
0	-3	9	-2	0	1		
0	0	7	-3	2	0		

The new matrix of A is reduced to echelon form after one stage.

Hence

det **A** = $\dfrac{(-1) \times 2 \times -3 \times 7}{1 \times 2}$ = 21

The (-1) appearing is due to the initial interchange of two rows of the matrix **A**.

By bvs,

$7y_3 = (-3, \ 2, \ 0)$

Thus, since det **A** = 21, we have;

$7(21)y_3 = 21(-3, \quad 2. \quad 0)$

$21y_3 = (-9, \quad 6, \quad 0)$

By further bvs;

$y_2 = \dfrac{1}{21}(-13, \ 18, -7)$

$y_1 = \dfrac{1}{21}(6, \ 3, \ 0)$

Therefore

$$A^{-1} = \frac{1}{21}\begin{pmatrix} 6 & 3 & 0 \\ -13 & 18 & -7 \\ -9 & 6 & 0 \end{pmatrix}$$

det **A** = 21

The solution of the system is given by **x** = **A**⁻¹ **b**

Hence

$$\begin{pmatrix} x_1 \\ x_2 \\ x_3 \end{pmatrix} = \frac{1}{21}\begin{pmatrix} 6 & 3 & 0 \\ -13 & 18 & -7 \\ -9 & 6 & 0 \end{pmatrix}\begin{pmatrix} 4 \\ 5 \\ 6 \end{pmatrix}$$

i.e.

$$\begin{pmatrix} x_1 \\ x_2 \\ x_3 \end{pmatrix} = \frac{1}{21}\begin{pmatrix} 39 \\ -4 \\ -6 \end{pmatrix}$$

This solution is exact as a rational vector. But if all the components of the solution vector are exactly divisible by the determinant without approximations, we should carry out the division.

Example 3 (b)

Use Jibunoh's method to solve Example 3 (a) without finding the inverse of the matrix.

If the inverse of the equation matrix is not required, the determinant of the matrix may be found and the system solved directly as follows.

2	0	-1	4	2	3
4	-3	7	6	(1)	
3	0	2	5		(2)
2	0	-1	4		
0	-3	9	-2		
0	0	7	-2		

Observe that due to the interchange of the 2nd and 3rd rows of the matrix, the 2nd and 3rd components of the RHS vector are also interchanged.

The RHS vector is usually not tampered with, when solving by going through the inverse of the equation matrix, as in part (a) of this example. From the echelon matrix above, we obtain;

$$\det \mathbf{A} = \frac{(-1)\times 2 - 3\times 7}{1\times 2} = 21, \text{ as before}$$

Since we are solving the equation directly without passing through the inverse of the equation matrix we use x_1, x_2, x_3 directly, in the backward substitutions.

Hence by bvs, we have

$$7x_3 = -2$$

Since det $\mathbf{A}$ = 21, we obtain

$$7(21)x_3 = -2(21) = -42$$

$x_3 = \frac{-6}{21}$, and by further bvs, $x_2 = -\frac{4}{21}$, $x_1 = \frac{39}{21}$

Hence

$$\begin{pmatrix} x_1 \\ x_2 \\ x_3 \end{pmatrix} = \frac{1}{21}\begin{pmatrix} 39 \\ -4 \\ -6 \end{pmatrix}, \quad \text{as before}$$

Solving directly as above without finding the inverse of the equation matrix may be faster, but there is advantage in obtaining the inverse, especially if new solutions are to be found later by varying the RHS of the system. In such a case, the matrix inverse is simply multiplied by the new RHS vector to obtain the new solutions.

Example 4

By using Jibunoh's method, find the determinant and inverse of the 4 x 4 matrix, taken from [3]

$$A = \begin{pmatrix} 5 & 4 & 2 & 1 \\ 2 & 3 & 1 & -2 \\ -5 & -7 & -3 & 9 \\ 1 & -2 & -1 & 4 \end{pmatrix}$$

Hence solve the system of equations

$$5x_1 + 4x_2 + 2x_3 + x_4 = 10$$
$$2x_1 + 3x_2 + x_3 - 2x_4 = 15$$
$$-5x_1 - 7x_2 - 3x_3 + 9x_4 = -10$$
$$x_1 - 2x_2 - x_3 + 4x_4 = 12$$

We proceed with the following reductions;

5	4	2	1	1	0	0	0	2	1	1
2	3	1	-2	0	1	0	0	(5)		
-5	-7	-3	9	0	0	1	0		(1)	
1	-2	-1	4	0	0	0	1			(5)
5	4	2	1	1	0	0	0			
0	7	1	-12	-2	5	0	0	3	2	
0	-3	-1	10	1	0	1	0	(7)		
0	-14	-7	19	-1	0	0	5		(1)	
5	4	2	1	1	0	0	0			
0	7	1	-12	-2	5	0	0			
0	0	-4	34	1	15	7	0	5		
0	0	-5	-5	-5	10	0	5	(4)		

5	4	2	1	1	0	0	0
0	7	1	-12	-2	5	0	0
0	0	-4	34	1	15	7	0
0	0	0	-190	-25	-35	-35	20

$$\det \mathbf{A} = \frac{5 \times 7 \times -4 \times -190}{5 \times 1 \times 5 \times 7 \times 1 \times 4} = 38$$

We now have

$$\begin{pmatrix} 5 & 4 & 2 & 1 \\ 0 & 7 & 1 & -12 \\ 0 & 0 & -4 & 34 \\ 0 & 0 & 0 & -190 \end{pmatrix} \begin{pmatrix} y_1 \\ y_2 \\ y_3 \\ y_4 \end{pmatrix} = \begin{pmatrix} 1 & 0 & 0 & 0 \\ -2 & 5 & 0 & 0 \\ 1 & 15 & 7 & 0 \\ -25 & -35 & -35 & 20 \end{pmatrix}$$

Since det **A** = 38, then by bvs;

$-190(38)y_4 = 38(-25, -35, -35, 20)$

$38y_4 = (5, 7, 7, -4)$

$y_4 = \frac{1}{38}(5, 7, 7, -4)$

By further bvs;

$y_3 = \frac{1}{38}(33, -83, -7, -34)$

$y_2 = \frac{1}{38}(-7, 51, 13, -2)$

$y_1 = \frac{1}{38}(-1, -9, -9, 16)$

Hence

$$\mathbf{A}^{-1} = \frac{1}{38} \begin{pmatrix} -1 & -9 & -9 & 16 \\ -7 & 51 & 13 & -2 \\ 33 & -83 & -7 & -34 \\ 5 & 7 & 7 & -4 \end{pmatrix}$$

Solving the given system, we obtain

$$\begin{pmatrix} x_1 \\ x_2 \\ x_3 \\ x_4 \end{pmatrix} = A^{-1} \begin{pmatrix} 10 \\ 15 \\ -10 \\ 12 \end{pmatrix} = \frac{1}{38} \begin{pmatrix} 137 \\ 541 \\ -1253 \\ 37 \end{pmatrix}$$

Example 5

Use Jibunoh's method to obtain the determinant and the inverse of the 6 x 6 matrix given by;

$$A = \begin{pmatrix} 0 & 2 & 1 & 4 & -1 & 3 \\ 1 & 2 & -1 & 3 & 4 & 0 \\ 0 & 1 & 1 & -1 & 2 & -1 \\ 2 & 3 & -4 & 2 & 0 & 5 \\ 1 & 1 & 1 & 3 & 0 & 2 \\ -1 & -1 & 2 & -1 & 2 & 0 \end{pmatrix}$$

This square matrix of order 6, is taken from Burden and Faires [1]. We interchange the first and second rows to have a number other than zero, to lead the pivot row, since by theory [2], a_{11} of **A**, must be non-zero. This interchange of two rows, of course, means that the eventual determinant shall be multiplied by (-1). We therefore proceed with the following reductions;

1	2	-1	3	4	0	0	1	0	0	0	0	2	1	1
0	2	1	4	-1	3	1	0	0	0	0	0			
0	1	1	-1	2	-1	0	0	1	0	0	0			
2	3	-4	2	0	5	0	0	0	1	0	0	(1)		
1	1	1	3	0	2	0	0	0	0	1	0		(1)	
-1	-1	2	-1	2	0	0	0	0	0	0	1			(1)

1	2	-1	3	4	0	0	1	0	0	0	0				
0	2	1	4	-1	3	1	0	0	0	0	0	1	1	1	1
0	1	1	-1	2	-1	0	0	1	0	0	0	(2)			
0	-1	-2	-4	-8	5	0	-2	0	1	0	0		(2)		
0	-1	2	0	-4	2	0	-1	0	0	1	0			(2)	
0	1	1	2	6	0	0	1	0	0	0	1				(2)

1	2	-1	3	4	0	0	1	0	0	0	0			
0	2	1	4	-1	3	1	0	0	0	0	0			
0	0	1	-6	5	-5	-1	0	2	0	0	0	3	5	1
0	0	-3	-4	-17	13	1	-4	0	2	0	0	(1)		
0	0	5	4	-9	7	1	-2	0	0	2	0		(1)	
0	0	1	0	13	-3	-1	2	0	0	0	2			(1)

1	2	-1	3	4	0	0	1	0	0	0	0		
0	2	1	4	-1	3	1	0	0	0	0	0		
0	0	1	-6	5	-5	-1	0	2	0	0	0		
0	0	0	-22	-2	-2	-2	-4	6	2	0	0	34	6
0	0	0	34	-34	32	6	-2	-10	0	2	0	(22)	
0	0	0	6	8	2	0	2	-2	0	0	2		(22)

1	2	-1	3	4	0	0	1	0	0	0	0	
0	2	1	4	-1	3	1	0	0	0	0	0	
0	0	1	-6	5	-5	-1	0	2	0	0	0	
0	0	0	-22	-2	-2	-2	-4	6	2	0	0	
0	0	0	0	-816	636	64	-180	-16	68	44	0	164
0	0	0	0	164	32	-12	20	-8	12	0	44	(816)

1	2	-1	3	4	0	0	1	0	0	0	0
0	2	1	4	-1	3	1	0	0	0	0	0
0	0	1	-6	5	-5	-1	0	2	0	0	0
0	0	0	-22	-2	-2	-2	-4	6	2	0	0
0	0	0	0	-816	636	64	-180	-16	68	44	0
0	0	0	0	0	130416	704	-13200	-9152	20944	7216	35904

$$\det \mathbf{A} = \frac{(-1) \times 1 \times 2 \times 1 \times -22 \times -816 \times 130416}{1 \times 1 \times 1 \times 2 \times 2 \times 2 \times 2 \times 1 \times 1 \times 1 \times 22 \times 22 \times 816} = -741$$

The (-1) appearing is due to the initial interchange of the first and second rows of the matrix **A**.

By bvs, and since det **A** = -741, we have;

$130416(-741)y_6 = -741(704, \ -13200, \ -9152, \ 20944, \ 7216, \ 35904)$

$741 y_6 = \frac{1}{176}(704, \ -13200, \ -9152, \ 20944, \ 7216, \ 35904)$

$y_6 = \frac{1}{741}(4, \ -75, \ -52, \ 119, \ 41, \ 204)$

By further bvs;

$y_5 = \frac{1}{741}(-55, \ 105, \ -26, \ 31, \ -8, \ 159)$

$y_4 = \frac{1}{741}(72, \ 132, \ -195, \ -81, \ -3, \ -33)$

$y_3 = \frac{1}{741}(-14, \ -108, \ 182, \ -46, \ 227, \ 27)$

$y_2 = \frac{1}{741}(200, \ -45, \ 364, \ 22, \ -173, \ -174)$

$y_1 = \frac{1}{741}(-410, \ -93, \ 143, \ 29, \ 614, \ -162)$

Hence

$$\mathbf{A}^{-1} = \frac{1}{741} \begin{pmatrix} -410 & -93 & 143 & 29 & 614 & -162 \\ 200 & -45 & 364 & 22 & -173 & -174 \\ -14 & -108 & 182 & -46 & 227 & 27 \\ 72 & 132 & -195 & -81 & -3 & -33 \\ -55 & 105 & -26 & 31 & -8 & 159 \\ 4 & -75 & -52 & 119 & 41 & 204 \end{pmatrix}$$

Example 6

By using Jibunoh's method, obtain the determinant and inverse of the matrix

$$A = \begin{pmatrix} \frac{1}{4} & \frac{1}{5} & \frac{1}{6} \\ \frac{1}{3} & \frac{1}{4} & \frac{1}{5} \\ \frac{1}{2} & 1 & 2 \end{pmatrix}$$

We simplify the procedure, as explained in section 3, by converting A to A* with integral entries. This is achieved by multiplying the first row by $\beta_1 = 60$, the second row by $\beta_2 = 60$ and third row by $\beta_3 = 2$.

Hence by (3.2),

$$\begin{pmatrix} 60 & 0 & 0 \\ 0 & 60 & 0 \\ 0 & 0 & 2 \end{pmatrix} \begin{pmatrix} \frac{1}{4} & \frac{1}{5} & \frac{1}{6} \\ \frac{1}{3} & \frac{1}{4} & \frac{1}{5} \\ \frac{1}{2} & 1 & 2 \end{pmatrix} = A^*$$

which gives

$$A^* = \begin{pmatrix} 15 & 12 & 10 \\ 20 & 15 & 12 \\ 1 & 2 & 4 \end{pmatrix}$$

By reduction of A* to echelon form as in the previous examples, we obtain
detA* = -26
and

$$A^{*-1} = \frac{1}{26} \begin{pmatrix} -36 & 28 & 6 \\ 68 & -50 & -20 \\ -25 & 18 & 15 \end{pmatrix}$$

Then using (3.4);

$$\det A = \frac{\det A^*}{\beta_1 \beta_2 \beta_3} = \frac{-13}{3600}$$

Now;

$$A^{-1} = A^{*-1} \begin{pmatrix} \beta_1 & 0 & 0 \\ 0 & \beta_2 & 0 \\ 0 & 0 & \beta_3 \end{pmatrix}, \text{ by (3.5)}$$

Hence

$$A^{-1} = \frac{1}{26}\begin{pmatrix} -2160 & 1680 & 12 \\ 4080 & -3000 & -40 \\ -1500 & 1080 & 30 \end{pmatrix} = \frac{1}{13}\begin{pmatrix} -1080 & 840 & 6 \\ 2040 & -1500 & -20 \\ -750 & 540 & 15 \end{pmatrix}$$

Example 7

Use Jibunoh's method to find the determinant and inverse of the complex matrix given by

$$A = \begin{pmatrix} 3 & 2-i & 4+i \\ 2-i & 6 & i \\ 4+i & i & 3 \end{pmatrix}$$

Applying the simplification procedures in section 4, we proceed with the following reductions;

3	2-i	4+i	1	0	0	2-i	4+i
2-i	6	i	0	1	0	(3)	
4+i	i	3	0	0	1		(3)
3	2-i	4+i	1	0	0		
0	15+4i	-9+5i	-2+i	3	0	-9+5i	
0	-9+5i	-6-8i	-4-i	0	3	(15+4i)	
3	2-i	4+i	1	0	0		
0	15+4i	-9+5i	-2+i	3	0		
0	0	-114-54i	-69 – 12i	27 – 15i	45 + 12i		

$$\det A = \frac{3 \times (15 + 4i)(-114 - 54i)}{3^2 \times (15 + 4i)} = -38 - 18i$$

To find the inverse of A we proceed as follows:

By bvs and the simplification procedure explained in section 4, we write

$$(-114 - 54i)y_3 = \begin{pmatrix} -69 - 12i \\ 27 - 15i \\ 45 + 12i \end{pmatrix}^T$$

Therefore

$$y_3 = \frac{1}{114 + 54i} \begin{pmatrix} 69 + 12i \\ -27 + 15i \\ -45 - 12i \end{pmatrix}^T$$

By further bvs,

$$(15 + 4i)y_2 + (-9 + 5i)y_3 = (-2 + i,\ 3,\ 0)$$

i.e

$$(15+4i)y_2 + \frac{1}{114+54i}\begin{pmatrix} 69 & 12 \\ -27 & 15 \\ -45 & -12 \end{pmatrix}\begin{pmatrix} -9 & 5 \\ -5 & -9 \end{pmatrix} = \frac{1}{114+54i}\begin{pmatrix} -2 & 1 \\ 3 & 0 \\ 0 & 0 \end{pmatrix} \times \begin{pmatrix} 114 & 54 \\ -54 & 114 \end{pmatrix},$$

converting the RHS to have 114 + 54i as the denominator.

Then after simplification;

$$y_2 = \frac{1}{(15+4i)(114+54i)} \begin{pmatrix} 399 - 231i \\ 174 + 432i \\ -465 + 117i \end{pmatrix}^T$$

By further bvs;

$$3y_1 + (2-i)y_2 + (4+i)y_3 = (1,\ 0\ \ 0)$$

Therefore by substituting the known y_2 and y_3, we have;

$$3y_1 + \frac{1}{(15+4i)(114+54i)} \begin{pmatrix} 399 & -231 \\ 174 & 432 \\ -465 & 117 \end{pmatrix} \begin{pmatrix} 2 & -1 \\ 1 & 2 \end{pmatrix}$$

+

$$\frac{1}{(15+4i)(114+54i)} \begin{pmatrix} 69 & 12 \\ -27 & 15 \\ -45 & -12 \end{pmatrix} \begin{pmatrix} 4 & 1 \\ -1 & 4 \end{pmatrix} \times \begin{pmatrix} 15 & 4 \\ -4 & 15 \end{pmatrix}$$

$$= \frac{1}{(15+4i)(114+54i)} \begin{pmatrix} 1 & 0 \\ 0 & 0 \\ 0 & 0 \end{pmatrix} \times \begin{pmatrix} 15 & 4 \\ -4 & 15 \end{pmatrix} \times \begin{pmatrix} 114 & 54 \\ -54 & 114 \end{pmatrix}$$

reducing all terms on the RHS of $3y_1$ to the same denominator $(15+4i)(114+54i)$

After simplification;

$$y_1 = \frac{1}{3(15+4i)(114+54i)} \begin{pmatrix} -2565 - 684i \\ 1197 - 693i \\ 2961 + 1368i \end{pmatrix}^T$$

By bringing the known y_2 and y_3 to have the same denominator $3(15+4i)(114+54i)$, as y_1, we obtain;

$$y_2 = \frac{1}{3(15+4i)(114+54i)} \begin{pmatrix} 399 - 231i \\ 174 + 432i \\ -465 + 117i \end{pmatrix}^T \times 3$$

$$= \frac{1}{3(15+4i)(114+54i)} \begin{pmatrix} 1197 - 693i \\ 522 + 1296i \\ -1395 + 351i \end{pmatrix}^T$$

and

$$y_3 = \frac{1}{3(15+4i)(114+54i)} \begin{pmatrix} 69 & 12 \\ -27 & 15 \\ -45 & -12 \end{pmatrix} \times \begin{pmatrix} 15 & 4 \\ -4 & 15 \end{pmatrix} \times 3$$

$$= \frac{1}{3(15+4i)(114+54i)} \begin{pmatrix} 2961 & + & 1368i \\ -1395 & + & 351i \\ -1881 & - & 1080i \end{pmatrix}^T$$

Now 3(15 + 4i) (114 + 54i) = 4482 + 3798i

Therefore

$$A^{-1}(\text{Analytical}) = \frac{1}{4482+3798i} \begin{pmatrix} -2565-684i & 1197-693i & 2961+1368i \\ 1197-693i & 522+1296i & -1395+351i \\ 2961+1368i & -1395+351i & -1881-1080i \end{pmatrix}$$

By definition, this is the *Analytical inverse* of **A**.

To convert A^{-1} *Analytical* to A^{-1} *Empirical* we use the conjugate of 4482 + 3798i to obtain the new multiplicative factor $\frac{4482-3798i}{34513128}$

Then;

$$A^{-1} = \frac{1}{34513128} \begin{pmatrix} -2565-684i & 1197-693i & 2961+1368i \\ 1197-693i & 522+1296i & -1395+351i \\ 2961+1368i & -1395+351i & -1881-1080i \end{pmatrix} \times (4482-3798i)$$

This leads to

$$\frac{1}{34513128} \begin{pmatrix} (-2565,-684) & (1997,-693) & (2961,1368) \\ (1197,-693) & (522,1296) & (-1395,351) \\ (2961,1368) & (-1395,351) & (-1881,-1080) \end{pmatrix} \begin{pmatrix} 4482 & -3798 \\ 3798 & 4482 \end{pmatrix}$$

where each row-vector entry is multiplied by the matrix on the right handside to obtain the complex entries as usual. Then dividing the resultant complex entries by 162, the common factor, we obtain;

$$A^{-1} = \frac{1}{213044} \begin{pmatrix} -87001+41211i & 16870-47236i & 113993-31571i \\ 16870-47236i & 44826+23618i & -30366+42416i \\ 113993-31571i & -30366+42416i & -77361+14219i \end{pmatrix}$$

It follows that;

$$A^{-1}(\text{Empirical}) = \frac{1}{213044}\begin{pmatrix} -87001 & 16870 & 113993 \\ 16870 & 44826 & -30366 \\ 113993 & -30366 & -77361 \end{pmatrix} + \frac{1}{213044}\begin{pmatrix} 41211 & -47236 & -31571 \\ -47236 & 23618 & 42416 \\ -31571 & 42416 & 14219 \end{pmatrix}i$$

Example 8

By first finding the determinant and empirical inverse of the equation matrix, solve the complex system;

$(2-i)x_1 + (3+i)x_2 = 5-i$
$(2+i)x_2 + (4+3i)x_2 = 6+2i$

The equation matrix is;

$$A = \begin{pmatrix} 2-i & 3+i \\ 2+i & 4+3i \end{pmatrix}$$

By Jibunoh's method we proceed with the usual reductions, using the simplification procedures, as in Example 7.

2-i	3+i	1	0	2+i
2+i	4+3i	0	1	(2-i)
2-i	3+i	1	0	
0	6-3i	-2-i	2-i	

$$\det A = \frac{(2-i)(6-3i)}{(2-i)} = 6-3i$$

By bvs

$$y_2 = \frac{1}{6-3i}(-2-i,\ 2-i)$$

Therefore

$$y_2 = \frac{1}{(6-3i)}\begin{pmatrix} -2-i \\ 2-i \end{pmatrix}^T$$

By further bvs;

(2- i)y₁ + (3+ i)y₂ = (1, 0)

i.e

$$(2-i)y_1 + \frac{1}{(6-3i)}\begin{pmatrix} -2 & -1 \\ 2 & -1 \end{pmatrix}\begin{pmatrix} 3 & 1 \\ -1 & 3 \end{pmatrix} = \frac{1}{(6-3i)}\begin{pmatrix} 1 & 0 \\ 0 & 0 \end{pmatrix} \times \begin{pmatrix} 6 & -3 \\ 3 & 6 \end{pmatrix}$$

Thus

$$(2-i)y_1 = \frac{1}{(6-3i)}\begin{pmatrix} 11+2i \\ -7+i \end{pmatrix}^T$$

Then

$$y_1 = \frac{1}{(2-i)(6-3i)}\begin{pmatrix} 11+2i \\ -7+i \end{pmatrix}^T$$

Bringing y₂ to the same denominator as y₁, we have;

$$y_2 = \frac{1}{(2-i)(6-3i)}\begin{pmatrix} -2 & -1 \\ 2 & -1 \end{pmatrix} \times \begin{pmatrix} 2 & -1 \\ 1 & 2 \end{pmatrix}$$

or

$$y_2 = \frac{1}{(2-i)(6-3i)}\begin{pmatrix} -5 \\ 3-4i \end{pmatrix}^T$$

Therefore;

$$A^{-1}(\text{Analytical}) = \frac{1}{9-12i}\begin{pmatrix} 11+2i & -7+i \\ -5 & 3-4i \end{pmatrix}$$

Now

$$\frac{9+12i}{225}\begin{pmatrix} 11+2i & -7+i \\ -5 & 3-4i \end{pmatrix} = A^{-1}(\text{Empirical}).$$

Hence

$$A^{-1}(\text{Empirical}) = \frac{1}{225}\begin{pmatrix} 75 & -75 \\ -45 & 75 \end{pmatrix} + \frac{1}{225}\begin{pmatrix} 150 & -75 \\ -60 & 0 \end{pmatrix}i$$

or

$$A^{-1}(\text{Empirical}) = \frac{1}{15}\begin{pmatrix} 5 & -5 \\ -3 & 5 \end{pmatrix} + \frac{1}{15}\begin{pmatrix} 10 & -5 \\ -4 & 0 \end{pmatrix}i$$

The RHS of the given complex system is

$$b = \begin{pmatrix} 5 \\ 6 \end{pmatrix} + \begin{pmatrix} -1 \\ 2 \end{pmatrix} i$$

Thus applying (5.3) we obtain

$$\mathbf{x} = \frac{1}{15}(15 + 5i, \quad 11 - 7i)^T$$

as the solution of the system.

Observe in general that the solution **x** is easily obtained for any k x k complex system once the *Empirical Inverse* is found.

7. Conclusion

The procedure of obtaining the inverse of any square matrix or solving a k × k system of linear equations via Jibunoh's determinants [2], is obviously easier than the traditional application of cofactors, Gaussian elimination, or iterative methods in the literature. The procedure gives the exact inverse of a real or complex matrix and exact solutions of systems. In the inversion of a complex matrix, we obtain what is defined as either the Analytical or Empirical inverse. The entries of all inverse matrices are mainly integers (without the multiplying factors). Hence the procedure is superior, in accuracy, to computer solutions which are usually decimal representations.

Even though the present application of Jibunoh's method is manual, it can evaluate with computational ease any k x k matrix or system, as $k \to \infty$. We may, however, note that in finding determinants, a computer program of Jibunoh's method is possible. But for matrix inversions, the snag is whether, as in Jibunoh's method, the computer program can retain the entries of the inverse matrices as integers or rational numbers, without decimals. This aspect needs investigation, especially in this age of automation.

8. EXERCISES

The reader is encouraged to attempt the following exercises using Jibunoh's method.

(I) Let $\mathbf{A} = \begin{pmatrix} 7 & 2 & 0 \\ 3 & 5 & -1 \\ 0 & -5 & -6 \end{pmatrix}$

Show that

(a) det $\mathbf{A}$ = -209

and

(b) $A^{-1} = \dfrac{1}{209} \begin{pmatrix} 35 & -12 & 2 \\ -18 & 42 & -7 \\ 15 & -35 & -29 \end{pmatrix}$

(II) Solve the system of equations, (1.1), which is discussed in the Introduction; i.e.

$0.003x_1 + 59.14x_2 = 59.17$

$5.291x_1 - 6.13x_2 = 46.78$

Answer

$\mathbf{x} = (10, 1)^T$

(III) Let

$$\mathbf{A} = \begin{pmatrix} \dfrac{1}{2} & -1 & -\dfrac{1}{3} & \dfrac{1}{6} \\ \dfrac{3}{4} & \dfrac{1}{2} & -1 & \dfrac{1}{4} \\ 1 & -4 & 1 & 2 \\ \dfrac{1}{2} & \dfrac{3}{4} & \dfrac{1}{2} & \dfrac{1}{4} \end{pmatrix}$$

By converting the matrix A to a matrix A* with integral entries, find det A and A^{-1} using Jibunoh's method. <u>Hint:</u> Use the LCMs of the denominators of fractions to convert.

Answer

(a) det A = $-\dfrac{97}{48}$

(b) $A^{-1} = \dfrac{1}{194}\begin{pmatrix} 342 & -52 & -50 & 224 \\ -126 & 60 & -2 & 40 \\ 78 & -148 & -8 & 160 \\ -462 & 220 & 122 & -112 \end{pmatrix} = \dfrac{1}{97}\begin{pmatrix} 171 & -26 & -25 & 112 \\ -63 & 30 & -1 & 20 \\ 39 & -74 & -4 & 80 \\ -231 & 110 & 61 & -56 \end{pmatrix}$

(IV) A 2 x 2 complex matrix is given by

$$A = \begin{pmatrix} 2 & 3-i \\ 2+i & 4+3i \end{pmatrix}$$

Using Jibunoh's method, find
(a) det A
(b) A^{-1} (Analytical)
(c) A^{-1} (Empirical)

Answer
(a) det A = 1 + 5i

(b) A^{-1} (Analytical) = $\dfrac{1}{1+5i}\begin{pmatrix} 4+3i & -3+i \\ -2-i & 2 \end{pmatrix}$

(c) A^{-1} (Empirical) = $\dfrac{1}{26}\begin{pmatrix} 19 & 2 \\ -7 & 2 \end{pmatrix} + \dfrac{1}{26}\begin{pmatrix} -17 & 16 \\ 9 & -10 \end{pmatrix}i$

(V) Solve the complex linear system

$$2x_1 + (3-i)x_2 = 4-i$$
$$(2+i)x_1 + (4+3i)x_2 = 6+i$$

<u>Hint:</u> Observe that the equation matrix is the matrix A given in Exercise **(IV)** above, which has the empirical inverse given by (c).

Answer

$$\mathbf{x} = \frac{1}{26}(55+11i\ ,\ 3-15i)^T$$

(VI) Find the determinant of the equation matrix A, the inverse of A, and the solution of the system taken from [1], p383.

$$6x_1 + 2x_2 + x_3 - x_4 = 0$$
$$2x_1 + 4x_2 + x_3 = 7$$
$$x_1 + x_2 + 4x_3 - x_4 = -1$$
$$-x_1 - x_3 + 3x_4 = -2$$

Answer

(a) det A = 191

(b) $A^{-1} = \dfrac{1}{191} \begin{pmatrix} 41 & -20 & -2 & 13 \\ -20 & 61 & -13 & -11 \\ -2 & -13 & 56 & 18 \\ 13 & -11 & 18 & 74 \end{pmatrix}$

(c) $\mathbf{x} = \dfrac{1}{191}(-164,\ 462,\ -183,\ -243)^T$

(VII) Let

$$10x_1 - x_2 + 2x_3 = 6$$
$$-x_1 + 11x_2 - x_3 + 3x_4 = 25$$
$$2x_1 - x_2 + 10x_3 - x_4 = -11$$
$$3x_2 - x_3 + 8x_4 = 15$$

as obtained from [1] p. 407

The system above was solved by the Jacobi iterative method. The tenth iterate could not obtain the exact solution. Let A be the equation matrix. Use Jibunoh's method to find det A, A^{-1} and the solution of the system.

Answer

(a) det A = 7395

(b) $A^{-1} = \dfrac{1}{7395} \begin{pmatrix} 777 & 69 & -153 & -45 \\ 69 & 758 & 34 & -280 \\ -153 & 34 & 782 & 85 \\ -45 & -280 & 85 & 1040 \end{pmatrix}$

(c) $\mathbf{x} = (1, \ 2, \ -1, \ 1)^T$

(VIII) Consider the tri-diagonal system of equations;

$2x_1 - x_2 \qquad\qquad\qquad = 1$
$-x_1 + 2x_2 - x_3 \qquad\quad = 0$
$\qquad -x_2 + 2x_3 - x_4 = 0$
$\qquad\qquad -x_3 + 2x_4 = 1$

which was solved in [1] p 381, using the Crout Factorization Algorithm. Denote the equation matrix by A. By applying Jibunoh's method, find

(a) det A
(b) A^{-1}
(c) The solution $\mathbf{x}$

Answer

(a) det A = 5

(b) $A^{-1} = \dfrac{1}{5} \begin{pmatrix} 4 & 3 & 2 & 1 \\ 3 & 6 & 4 & 2 \\ 2 & 4 & 6 & 3 \\ 1 & 2 & 3 & 4 \end{pmatrix}$

(c) $\mathbf{x} = (1, 1, 1, 1)^T$

(IX) The following linear system is obtained from Burden and Faires [1] p.383.

$$\begin{aligned} 0.5x_1 + 0.25x_2 &= 0.35 \\ 0.35x_1 + 0.8x_2 + 0.4x_3 &= 0.77 \\ 0.25x_2 + x_3 + 0.5x_4 &= -0.5 \\ x_3 - 2x_4 &= -2.25 \end{aligned}$$

Let A be the equation matrix. Transform the entries of A to integers by using
$$\beta_1 = \beta_2 = \beta_3 = \beta_4 = 100$$
Hence or otherwise show that

(a) det A = -0.68125

(b) $A^{-1} = \dfrac{1}{68125} \begin{pmatrix} 180000 & -62500 & 20000 & 5000 \\ -87500 & 125000 & -40000 & -10000 \\ 17500 & -25000 & 62500 & 15625 \\ 8750 & -12500 & 31250 & -26250 \end{pmatrix}$

and

(c) $\mathbf{x} = \dfrac{1}{68125} (-6375, \ 108125, \ -79531.25, \ 36875)^T$

(X) A complex system is given by

$$\begin{pmatrix} 1 & 1+i & 2i \\ 1-i & 4 & 2-3i \\ -2i & 2+3i & 7 \end{pmatrix} \begin{pmatrix} x_1 \\ x_2 \\ x_3 \end{pmatrix} = \begin{pmatrix} 2-i \\ 3+4i \\ 5-2i \end{pmatrix}$$

where the equation matrix A is Hermitian

By applying Jibunoh's method, show that

(a) det A = -19

(b) A^{-1} (Empirical) $= \dfrac{1}{19}\begin{pmatrix} -15 & 13 & -5 \\ 13 & -3 & 0 \\ -5 & 0 & -2 \end{pmatrix} + \dfrac{1}{19}\begin{pmatrix} 0 & 3 & 9 \\ -3 & 0 & -5 \\ -9 & 5 & 0 \end{pmatrix}i$

and

(c) the solution $\mathbf{x} = \dfrac{1}{19}(-10+131i,\ 4-56i,\ -49+6i)^T$

(XI) In Exercise (X) above, let the real and imaginary parts of A^{-1}(Empirical) be denoted by V_1 and V_2 respectively and those for the matrix A be denoted by A_1 and A_2, such that

A^{-1}(Empirical) $= V_1 + iV_2$
and
$A = A_1 + iA_2$
Verify equations (4.10), ie.
$A_1 V_1 - A_2 V_2 = I$
$A_2 V_1 + A_1 V_2 = 0$

(XII) Let a 5 x 5 matrix be given by

$$A = \begin{pmatrix} 2 & 4 & 1 & 6 & 7 \\ 3 & 6 & 2 & 4 & 2 \\ 4 & -7 & 2 & 5 & -1 \\ -5 & 4 & 1 & 3 & 4 \\ 0 & 2 & 1 & 4 & 6 \end{pmatrix}$$

Use Jibunoh's method to find det A and A^{-1} Hence solve the system of equations;

$$\begin{aligned}
2x_1 + 4x_2 + x_3 + 6x_4 + 7x_5 &= 5 \\
3x_1 + 6x_2 + 2x_3 + 4x_4 + 2x_5 &= 16 \\
4x_1 - 7x_2 + 2x_3 + 5x_4 - x_5 &= -12 \\
-5x_1 + 4x_2 + x_3 + 3x_4 + 4x_5 &= 20 \\
2x_2 + x_3 + 4x_4 + 6x_5 &= -8
\end{aligned}$$

Answer

(a) det A = 1505

(b) $A^{-1} = \dfrac{1}{1505}\begin{pmatrix} 8 & 93 & -30 & -266 & 132 \\ 123 & 113 & -85 & 49 & -228 \\ -1624 & 686 & 70 & -182 & 1799 \\ 760 & -195 & 160 & 315 & -1005 \\ -277 & -22 & -90 & -196 & 697 \end{pmatrix}$

(c) $\mathbf{x} = \dfrac{1}{1505}(-4488,\ 6247,\ -16016,\ 13100,\ -10153)^T$

(XIII) The following system was solved in [1] p. 432 by an iterative technique.

$0.04x_1 + 0.01x_2 - 0.01x_3 = 0.06$
$0.2x_1 + 0.5x_2 - 0.2x_3 = 0.3$
$x_1 + 2x_2 + 4x_3 = 11$

Using Jibunoh's method, obtain det A where A is the equation matrix. Hence find A^{-1} and the solution of the system

Answer

(a) det A = 0.087

(b) $A^{-1} = \dfrac{1}{87}\begin{pmatrix} 2400 & -60 & 3 \\ -1000 & 170 & 6 \\ -100 & -70 & 18 \end{pmatrix}$

(c) $\mathbf{x} = \dfrac{1}{87}(159,\ 57,\ 171)^T$

The solution is exact by Jibunoh's method. The iterative technique in [1] obtained the solution approximately as
$\mathbf{x} = (1.8,\ 0.64,\ 1.9)^T$

(XIV) Let a 4 x 4 matrix be given by

$$A = \begin{pmatrix} -9 & 2 & 3 & 4 \\ 1 & -8 & 3 & 4 \\ 1 & 2 & -7 & 4 \\ 1 & 2 & 3 & -6 \end{pmatrix}$$

Use Jibunoh's method to show that det **A** = 0 and hence that the matrix is not invertible.

(XV)

The matrix $\mathbf{A} = \begin{pmatrix} 1 & 1+i & 2i & 3-i \\ 1-i & 4 & 2-3i & -5+2i \\ -2i & 2+3i & 7 & i \\ 3+2i & 1+i & 1-i & 2+5i \end{pmatrix}$

is a 4 x 4 complex matrix

Show that;
(a) det A = -(454+269i)
(b) A^{-1} (Analytical) =

$$\frac{1}{8626 + 5111i} \begin{pmatrix} -931 - 1938i & -57 + 969i & -1235 + 361i & 2375 - 399i \\ 3629 + 1995i & 1634 + 285i & 57 - 646i & -1064 + 532i \\ 437 - 2508i & -722 - 703i & 798 + 456i & 646 + 931i \\ 817 + 456i & -665 - 703i & 570 - 266i & 361 \end{pmatrix}$$

and

(c) A^{-1} (Empirical) =

$$\frac{1}{278477} \begin{pmatrix} -49684 & 12357 & -24399 & 51101 \\ 114959 & 43079 & -7784 & -17892 \\ -25066 & -27205 & 25524 & 28617 \\ 25978 & -25843 & 9854 & 8626 \end{pmatrix} + \frac{1}{278477} \begin{pmatrix} -33127 & 23961 & 26111 & -43159 \\ -3709 & -16324 & -16243 & 27776 \\ -66115 & -6576 & -402 & 13100 \\ -671 & -7383 & -14426 & -5111 \end{pmatrix} i$$

You may also find A^{-1} (Empirical) by automatic computation using (4.11).

REFERENCES
1. Burden, R. L and Faires, J.D, *Numerical Analysis (Fifth Edition)* PWS Publishing Co., Boston MA. (1993).
2. Jibunoh, C.C. *Jibunoh's Method for evaluating the determinant of an N x N matrix (A monograph on research discovery)*, Royal Pace Publications, Agbor, Delta State, Nigeria (2009).
3. Lipschutz, S. *Theory and Problems of Linear Algebra* Schaum's Outline Series, McGraw-Hill Book Company, New York (1968).

INDEX

Abstract, 1

Backward vector substitutions (bvs) – definition, 2

Complex system, 7, 26, 31, 33

Converting the fractional or decimal entries of matrices to integral entries, 4, 5, 21, 29, 33, 35

Entries of the equation matrix,
- complex entries, 5
- decimal entries, 4, 35
- fractional entries, 4, 21, 29
- integral entries, 3, 4

Formula for detA, by Jibunoh's method, 8

Imaginary part of the solution of a complex system, 8

Inverses of a matrix,
- real inverse, 1, 2, 8, 9, 11, 14, 17, 20, 21, 29, 30, 31, 32, 33, 35
- analytical inverse of a complex matrix, 1, 2, 7, 25, 27, 28, 30, 36
- empirical inverse of a complex matrix, 1, 2, 7, 25, 26, 27, 28, 30, 34, 36

Jibunoh's method, 2, 4, 8, 10, 11, 14, 16, 18, 21, 22, 26, 28, 29, 30, 32, 33, 35, 36

Jibunoh's determinants, iv, 1, 28

Matrix,
- complex matrix, 1, 5, 6, 7, 22, 36
- diagonal matrix, 5
- echelon form (matrix), 3, 6, 8, 10, 13, 17, 19, 21, 22, 26
- equation matrix, 1, 2, 26, 31, 32, 35
- hermitian matrix, 33
- lower triangular matrix, 3
- upper triangular matrix, 3

Microsoft Excel Package, 7

Pivot strategy or row, 1, 18

Rational Vector, 14

Real part of the solution of a complex system, 8

Row interchanges in matrix inversion, 12

Simplification procedures,
- when the entries of the matrix are complex numbers, 5, 6
- when the entries of the matrix are decimal numbers, 4, 5
- when the entries of the matrix are fractions, 4, 21
- when the complex matrix is large or has decimal entries, 6, 7

Solving a system directly without the inverse of the equation matrix, 14, 15

Solutions of complex systems, 7, 8, 26, 28

Solutions of real systems, 1, 2, 3, 4, 11, 16, 29, 31, 32, 33, 34, 35

Solution of a system by Jibunoh's method and bvs, 2, 3, 4

Tri-diagonal system, 32, 33